# CLEAN ENERGY
# COMMON SENSE

# CLEAN ENERGY
# COMMON SENSE

*An American Call to Action on*
*Global Climate Change*

## FRANCES BEINECKE
## WITH BOB DEANS

ROWMAN & LITTLEFIELD PUBLISHERS, INC.
*Lanham • Boulder • New York • Toronto • Plymouth, UK*

Published by Rowman & Littlefield Publishers, Inc.
A wholly owned subsidiary of The Rowman & Littlefield
Publishing Group, Inc.
4501 Forbes Boulevard, Suite 200, Lanham, Maryland 20706
http://www.rowmanlittlefield.com

Estover Road, Plymouth PL6 7PY, United Kingdom

Library of Congress Control Number: 2009938213

ISBN 978-1-4422-0317-4 (pbk : alk paper) —
ISBN 978-1-4422-0318-1 (electronic)

∞ ™ The paper used in this publication meets the minimum
requirements of American National Standard for Information
Sciences—Permanence of Paper for Printed Library Materials,
ANSI/NISO Z39.48-1992.

Printed in the United States of America

*To the staff and board of the Natural Resources Defense Council—past and present—for their commitment to our mission to safeguard the Earth.*

# Contents

# FOREWORD

## By Robert Redford

In January of 1776, Philadelphia essayist Thomas Paine published a 47-page pamphlet that changed the world. Within three months, *Common Sense* had sold 150,000 copies. By July of that year, the national conversation charged by Paine's work culminated in the *Declaration of Independence*.

Frances Beinecke, president of the Natural Resources Defense Council, has penned a modern version of Paine's revolutionary classic, calling on us, as a nation, to rise to the challenge of global climate change, an environmental ill of astounding proportions, while there's still time to act.

There's a way for each of us to act, as well, and that is to press Congress to pass clean energy legislation that will help our nation generate jobs, reduce our dangerous reliance on foreign oil, and create a healthier planet for ourselves and our children. Time is of such essence, Beinecke shows, that every American of conscience must be engaged. Reading this book is an essential first step.

"A long habit of not thinking a thing wrong gives it a superficial appearance of being right," Paine wrote in his poignant appeal for American action, going on to signal a people's yearning for liberty as "the cause of all mankind."

Such a cause is what we face today in the need to confront climate change, a widening scourge that threatens us all.

After decades of study and observation worldwide, the evidence before us is overwhelming. Temperatures are rising. Ice caps are melting. Deserts are spreading. Majestic landscapes from Montana's glaciers to Utah's Red Rocks—places I used to view as enduring symbols of American grit—are in decline.

Storms are growing more vicious, droughts more intense. Huge swaths of the Earth are becoming hostile to life in its diverse and interwoven forms. Hardpressed people in some places have quit the lands of their forebears to join a restless new breed of climate refugees, as competition heats to conflict over the growing scarcity of living space, water, and food.

And this may be just the beginning.

What is at stake here is nothing short of staggering.

It stretches far beyond partisan politics and election-year cycles and reaches into the future. What kind of Earth will we leave our children and their children? Will we bequeath them only photographs to learn what a glacier was or what a healthy forest looked like? Or will we actually build the clean energy systems that will hold our natural legacy intact for generations to come?

"We now know, all of us, that our world is warming," Beinecke writes. "We know it is harming us all. We know it is only going to get worse until we stand up and summon the will to stop it. And we know what it will take to do that. We must find the courage to begin."

There are people of good will who hear claims on both sides of the climate change debate and aren't sure what to believe. If that feels familiar, this little book is for you. There are those who want what's best for future generations, for our environment, and for our country, yet worry that our politicians might not get this right. If you're nodding your head, this book is for you. And there are people who care deeply about our common future and just don't see climate change as a priority. If that resonates, this book is for you too, perhaps especially for you.

In a clear and compelling tone, Beinecke draws from the most current and authoritative sources anywhere to lay out the case for American action against world climate change. She outlines solutions that can help get American workers back on their feet, strengthen our country, and set us on the path to a clean energy future. And she calls on each of us to take up paper and pen to urge Congress to act.

For the act of making our voices heard is the best of American politics. I have seen it work time and again—I have seen citizens, neighborhoods, entire communities carry the weight of truth to lawmakers. But to succeed, we must raise our voices loudly and fully. This is what Beinecke inspires us to do.

"I offer nothing more than simple facts, plain arguments and common sense," Paine wrote two centuries ago. That's precisely the approach Beinecke has taken in her stand against climate change. Simple facts, plain arguments, and common sense. It's all here in this concise yet powerful book.

# INTRODUCTION

In November 2007, Georgia Governor Sonny Perdue led a Capitol vigil to pray for rain after the state's worst drought in a century baked crops to dust and turned lakebeds to chalk. Two years later, he declared a state of emergency in seventeen counties, after record flooding took at least nine lives. President Obama called to lend his support. But before he picked up the phone, Obama had to break away from United Nations talks. The subject? Climate change.

"The threat from climate change is serious, it is urgent, and it is growing," Obama told his U.N. counterparts that day in September. We can curb that threat, he said, in ways that strengthen our economy and make our country more secure, and we must act now. "The time we have to reverse this tide is running out."

There's a connection between what happens to our planet and what happens in our lives. Arctic ice affects currents and weather. Widening deserts spread thirst and disease. Crops fail, people go hungry. Climate shifts, and we see epic escalations in the severity and frequency of the kinds of hurricanes that deluge our cities, the wildfires that ravage our land, and successive drought and flooding that can whipsaw states and entire regions in a spiraling vortex of unpredictable change.

It doesn't have to be this way. We can turn back climate change. We must summon the will to do so, before we run out of time.

I don't come to such strong words lightly. My position is based on three and a half decades of work with the Natural Resources Defense Council (NRDC), the best environmental advocacy organization, I believe, in the world. We are guided by our devotion to sound science, the rule of law, and the public interest, inspired as much by Abraham Lincoln as by Rachel Carson.

Starting out 40 years ago, the NRDC took a page out of the civil rights playbook, taking legal strategy employed to challenge school segregation and using it to fight for clean water and air. We will always remember that, when we stand up for the good of our planet, we walk in the shadow of great struggles others have led—those who bled for our freedom, labored that we might prosper, marched for justice, spoke out for equality, and still fight the good fight every day of their lives in our classrooms, our factories, our fields, and our farms to live up to our national purpose, as Lincoln put it, as the last best hope of Earth.

Our Earth needs us now, as never before.

As I travel this country and listen and learn, I hear the reasons people give for not taking action on climate change. Most come down to one of three things.

Some people regard climate change as a problem, but not a priority. Others wonder what taking action might mean for our economy or our place in the world. Still others worry our politicians might somehow get it wrong.

These are valid expressions of important concerns. They deserve a considered response. That is why I wrote this little book.

Climate change affects each and every one of us, each and every day. Taking action against it will strengthen our economy and make our country more secure. And the men and women we've elected to govern us have what it takes to lead.

I believe those things. That's why I support the clean energy legislation that's before our Congress.

Clean energy legislation can curb global climate change in a way that helps us to generate jobs, reduce our reliance on foreign oil, and create a healthier planet for ourselves and our children. It deserves, and it needs, our support.

This book is a call to action, one citizen's honest appeal. It is not a political treatise. It is not a partisan screed. Maybe that's because my politics on this are simple. I believe Democrats and Republicans alike have a real chance here to lead, to look to the future and show us the way to a better and brighter tomorrow.

I know in some quarters this issue elicits division and rancor and noise. I can't for the life of me understand why. If ever there were a threat to unite us, an opportunity we all might rally around, surely we can muster the collective will to prevail against a pall hanging over us all.

Global climate change is the single greatest environmental challenge of our generation. It is, though, far more than that. It is a humanitarian challenge. It is an economic challenge. It is a national security challenge. And it is a moral challenge, the great moral challenge of our time. We must rise to meet it, we must turn it around, or we will fail our forebears and our children.

As a young woman, I walked the towering forests of the Adirondack Mountains, listening to the wind high in the trees, swimming the clear and frigid waters of a wilderness lake, pondering the silent splendor of a moonlit summer night. Sometimes I imagined our founders, first wandering this richly blessed land. The pride and immensity of spirit. Possibility unfettered. Great dreams unbound.

They had come from a place where one man was king, and the rest found some way to serve. In this new land, they would find a new way. They would cut their own path in these woods.

And out of that passion and promise, a bold new idea took root. In this land, the people were sovereign. They would bow down before no king.

I still believe in that promise. In my heart, I still wander those woods. And I know, now more than ever, what we must do to keep faith with those dreams. We must stand up for what we believe in. We must speak out the truth we all know.

We know our old ways aren't working. We see clearly the damage they've done. There are gathering risks to our country if we keep heading down the same road. We must stop, check our compass, correct course, and move forward—while there's still time to act.

Frances Beinecke
October 2009
New York

# 1

# A Clear and
# Present Danger

It isn't often that the Pentagon, the U.S. Department of Commerce, the United Nations, the National Academy of Sciences, the European Union, the Environmental Protection Agency, the National Intelligence Council, and the Shell Oil Company agree on anything.

When it comes to the threat of global climate change, however, the evidence is in and the verdict is clear. These groups, and more, all concur. The world is warming at an alarming rate. Irreparable harm has already been done. Catastrophe looms . . . unless we take the necessary steps to avert it.

The Arctic ice cover is melting. In the past three decades, a third of it has gone. The deserts of Africa are spreading, fueling disputes and sometimes conflict over increasingly scarce clean water and food. Himalayan glaciers are in retreat, threatening rivers that one out of every five people on Earth depend on for water to drink and to grow crops.

Oceans are becoming more acidic, putting coral and shellfish at risk.

Ancient pines in Yellowstone are dying. Wildfires are burning eight million acres of American lands each year, twice as much as a decade ago, and the devastation is set to expand.

Global sea levels are on track to rise by two feet or more by this century's end, a change that would leave scores of coastal cities worldwide as vulnerable to flooding as New Orleans. The Everglades would go underwater. Ellis Island could be submerged.

Hurricanes, droughts, typhoons, and storms have become more severe. The world endures, on average, more than 400 weather-related disasters each year, more than twice as many as just two decades ago.

A 2003 heat wave killed 30,000 people in Europe, half of them in France, in what the chief scientific adviser to the British government called "the largest, single natural disaster on record in central Europe, as measured by human fatalities."

That's what hot weather does to some of the richest people on Earth. This is what it means for the poorest: "Green pasture land has turned to dust . . . water taps have run dry . . . communities fight over the last remaining pieces of fertile grazing land . . . children, dressed in little more than a sheet, are hiking 20 miles for a gallon of water" (The New York Times, September 8, 2009. Dateline: Lokori, Kenya).

From the Sahara to Central Asia, a belt of dry land growing ever thirstier has seen the first trickles of what Refugees International already fears will be wave upon wave of climate refugees on the move in the decades ahead. Closer to home for Americans, the summer of 2009 brought the worst drought in 70 years to Mexico, killing an estimated 50,000 cows and wiping out 17 million acres of cropland.

"The human impact of climate change is happening right now," former U.N. Secretary-General Kofi Annan declared last May, when the group he founded, the Global Humanitarian Forum, released a survey of the impact global warming is already having on the world's poor. "For those living on the brink of survival," he said, "climate change is a very real and dangerous hazard."

It isn't only low-income people who are threatened. In September 2009, the U.S. Central Intelligence Agency launched The Center on Climate Change and National Security. Its mission is to assess the national security risks posed to the United States by widening desertification; rising sea levels; population shifts and increasing competition for food, land, fresh water, and other natural resources of growing scarcity around the world; and to better evaluate "the effect environmental factors can have on political, economic and social stability overseas," the CIA stated in a September 25 press release. As CIA Director Leon Panetta put it in the release, "Decision makers need information and analysis on the effects climate change can have on security."

In 1989, then-President George H. W. Bush started asking questions about climate change. He had heard a lot of fussing and fighting. He wanted to know the truth.

Initiated by Bush, the U.S. Global Change Research Program was one of the most exhaustive undertakings in the annals of scientific inquiry. It was a 20-year mega-study commissioned by the U.S. Congress and conducted by 13 federal agencies over the course of four administrations—two of them Republican and two Democratic.

The results were made public last June in a 196-page report. This is how it begins:

"Observations show that warming of the climate is unequivocal."

Unequivocal: leaving no doubt, open to no misunderstanding. That's what unequivocal means.

"The global warming observed over the past 50

years is due primarily to human-induced emissions of heat-trapping gases," the report goes on. "These emissions come mainly from the burning of fossil fuels (coal, oil, and gas), with important contributions from the clearing of forests, agricultural practices and other activities."

As fossil fuel use has intensified over the past century, carbon dioxide emissions, the report continues, have already raised average temperatures by as much as 7 degrees Fahrenheit in parts of the United States, shortening winters, lengthening summers, and raising sea levels and water temperatures in ways that have begun to affect human health, farms, coastal areas, and water supplies.

Further warming is already inevitable, and, unless action is taken to limit carbon emissions from burning fossil fuels in cars, factories, and power plants, the world could warm another 11.5 degrees Fahrenheit by the end of the century, the report concludes, with potentially devastating effects that span the globe.

"Likely future changes for the United States and surrounding coastal waters include more intense hurricanes with related increases in wind, rain and storm surges," states the report. "These changes will affect human health, water supply, agriculture, coastal areas and many other aspects of society and the natural environment."

The project was led by the National Oceanic and Atmospheric Administration, the most authoritative source of weather and climate information anywhere in the world. Also on board: the Department of Defense, the National Science Foundation, the Department of State, and nine other U.S. government agencies. They augmented their own data with hundreds of peer-reviewed studies in 21 separate scientific disciplines ranging from ecological systems to atmospheric chemistry.

The work was audited by the National Academy of Sciences and then further scrutinized by an independent panel of experts from institutions like the National Center for Atmospheric Research, the Massachusetts Institute of Technology, and Texas A&M University.

If you're wondering whether climate change is real, these are good people to ask.

Their findings, moreover, echo similar conclusions reached separately by the United Nations, the 27 countries that make up the European Union, and, indeed, one of the largest and oldest oil companies in the world.

"The scientific evidence is now overwhelming," Shell Oil states on its website. "Climate change is a serious global threat, one that demands an urgent worldwide response."

That statement is nearly identical to language contained in a 2008 assessment by the British Treasury. Those views, in turn, are mirrored in dozens of independent assessments from groups as diverse as the National Intelligence Council and the World Health Organization.

"Climate change poses a clear and present danger to the United States of America," retired Navy Admiral Lee Gunn told the Senate Foreign Relations Committee in July. He warned of strategic bases falling prey to rising sea levels, weakened states falling into anarchy, and regional conflicts boiling over amid growing scarcities of food, water, and arable farmland.

"As the planet warms, rainfall patterns shift and extreme events such as droughts, floods, and forest fires become more frequent," the World Bank stressed in its World Development Report 2010, released in mid-September. "Millions in densely populated coastal areas and in island nations will lose their homes as the sea level rises. Poor people in Africa, Asia, and elsewhere face prospects of tragic crop failures; reduced agricultural productivity; and increased hunger, malnutrition, and disease."

Little wonder Kofi Annan calls global climate change "the greatest emerging humanitarian challenge of our time."

A towering edifice of global consensus has risen around immutable facts.

To argue against them, at this point, is the twenty-first-century equivalent of saying the Earth is flat. The accumulated science, the body of research, the physical evidence before our very eyes, is literally that definitive.

"The science has become more irrevocable than ever: climate change is happening. The evidence is all around us," United Nations Secretary-General Ban Ki-moon wrote in the report on climate change that the U.N. Environment Program released September 24, 2009. "The time for hesitation is over. The time to act is now."

This debate has been settled and scored. Now we all know the truth.

We know that our world is warming. We know it is harming us all. We know it is only going to get worse until we stand up and summon the will to stop it. And we know what it will take to do that. We must find the courage to begin.

"We know enough to act on climate change," the U.S. Climate Action Partnership, a coalition of corporate and environmental interests, wrote in "A Call for Action," a 2007 position paper. "Congress needs to enact legislation as quickly as possible."

# 2

# FIRE AND ICE

In mid-September, two German merchant ships made maritime history, becoming the first commercial vessels to transit the fabled Northeast Passage across the Arctic Ocean, a route that has enticed, and eluded, mariners since the dawn of ocean travel.

For at least the past 8,000 years, the Arctic has been too choked with ice for most ships to pass more than partway across. Attempts dating to the sixteenth century have ended in disaster; as recently as 1983 a Russian vessel was crushed by ice.

This year, though, as summer waned, the 12,700-ton *Beluga Fraternity* and its sister ship, the *Beluga Foresight*, cruised undaunted through the rapidly retreating ice cover along the top of the world.

"Small ice bergs, ice fields, and ice blocks were traversed without incident," Beluga Shipping GmbH proudly announced in a press release touting the landmark nautical achievement.

What made the shipping triumph possible, though,

is one of the more alarming consequences of global climate change. The northern ice cap is melting, the result of steady warming caused by the burning of fossil fuels. In the past 30 years, a third of the ice has vanished, an area roughly the size of the United States east of the Mississippi River. Soon, it may all be gone.

"There is a possibility of an ice-free Arctic Ocean for a short period in summer perhaps as early as 2015," states the April 2009 Arctic Marine Shipping Assessment, the result of a four-year study by the Arctic Council, an intergovernmental forum that gathers representatives from the United States, Canada, Russia, and five Scandinavian countries. "This would mean the disappearance of multiyear ice, as no sea ice would survive the summer melt season."

Arctic sea ice is a kind of speedometer showing how fast global warming is changing our world. We're traveling at breakneck speed.

"Our foot is stuck on the gas pedal; we have to pull it off," U.N. Secretary-General Ban Ki-moon told reporters in September, shortly after visiting the Arctic. "I traveled to the Polar ice rim to see for myself some of the visible impacts of climate change. . . . It is essential that we act on what the science tells us."

Just last year, I made my own trip to the Arctic, aboard the National Geographic ship *Endeavour*. Coursing through the Svalbard archipelago, halfway between the Arctic rim and the North Pole, we, too, were on a kind of pilgrimage to a place climatologists regard as ground zero for global warming. I stared in awe at an ice pack 12 stories high, only to learn how far it had receded in just the past decade.

There's a poignant beauty in the way nature speaks to us, choosing a wild, remote, and forbidding place to remind us how closely every part of our planet is linked to the choices we make in our lives. In this place of frozen majesty, far from the burning of fossil fuels, the Earth's icebox is melting, signaling, unmistakably, the time for us to do exactly as Ban says, and act on what we know.

To make clear what we know, more than 2,000 scientists from the 192 countries that belong to the United Nations pulled together the best climate work they could find in an effort overseen by the Intergovernmental Panel on Climate Change. The group issued its findings in 2007—and received the Nobel Peace Prize for its work.

"Warming of the climate system is unequivocal," the panel reported, "as is now evident from observations of increases in global average air and ocean temperatures, widespread melting of snow and ice and rising global average sea level."

During the past century, average global temperature rose about 1.3 degrees Fahrenheit to its present level

of around 57 degrees Fahrenheit, the panel concluded. That may not sound like a lot. The figure,
though, represents year-round average temperature
change. It reflects, in other words, a long-term trend
toward longer, hotter summers and shorter, warmer
winters. Those are important changes that are having
broad affects.

"These include increases in air and water temperatures, reduced frost days, increased frequency and in-
tensity of heavy downpours, a rise in sea level, and reduced snow cover, glaciers, permafrost and sea ice,"
according to the U.S. Global Change Research Program.

The rate of warming, moreover, is accelerating. The
15 hottest years on record have all occurred since
1991, and 11 of those have occurred since 1998, according to the U.N. World Meteorological Organization. Worldwide, 1998 was the hottest year on record,
with global temperatures soaring 1 degree Fahrenheit
above the 30-year average, as measured from 1961 to
1990.

We're in for additional warming of between 6.3 and
8.1 degrees Fahrenheit by the end of this century,
concludes a September 2009 report by the U.N. Environment Program, depending on how much is done
to reduce the pollution from burning fossil fuels.

That's more than double the level scientists agree could lead to environmental catastrophe.

Those temperatures, bear in mind, are global averages. Climate is changing at different rates in different parts of the world. In the United States, for instance, average temperatures have risen 1.6 degrees Fahrenheit just since 1951, with nearly all the warming coming after 1970, according to the U.S. Global Change Research Program. "Over the past 30 years, temperatures have risen faster in winter than in any other season, with average winter temperatures in the Midwest and northern Great Plains increasing more than 7 degrees F," the Global Change Research Program reports.

Some of the fastest warming on Earth is occurring in the Arctic, where temperatures have risen at double the global average. In Alaska and western Canada, temperatures are up 3.6 degrees Fahrenheit over just the past six decades.

As a result, Arctic freezing seasons are shorter, and summer melt times are longer.

Sea ice matters to the U.S. Navy. That's why a Defense Department meteorological satellite flies over the Arctic every 99 minutes, photographing the region day and night with sophisticated microwave imaging instruments that document the state of the ice with military precision.

Those satellite pictures show us what's happening to Arctic ice—it's vanishing before our eyes.

One way climatologists measure that is by taking stock of the ice cover at summer's end, when the ice plumbs its annual low. Winter freezing restores much of the ice cover in colder months. Over the past three decades, though, we've lost more in warm months than we've gained in cold.

As the summer of 1980 ended, there were 3 million square miles of Arctic sea ice. By the middle of September 2009, the area was 1.9 million square miles—the third lowest ever recorded.

The five lowest levels on record, in fact, have occurred over the past five years. Lowest of all was 2007, when Arctic ice covered just 1.7 million square miles, shocking experts who feared they were looking at the prospect of catastrophic collapse.

"It was stunning," said Claire Parkinson, a senior climatologist at the National Aeronautics and Space Administration's (NASA's) sprawling Goddard Space Flight Center in the Washington, D.C., suburb of Greenbelt, Maryland. "When we saw that huge retreat in 2007, one thought was definitely, 'Geez, maybe it's been thinning so much it's just ready to disappear altogether.'"

In the 31 years she's spent studying Arctic ice, Parkinson has labored largely in obscurity, tracking data that was as removed from public interest and picnic conversation as the North Pole.

"Now it's a hot topic," she said in an interview, catching herself on the irony. "Hot is not good."

It isn't only sea ice that's affected. Arctic glaciers—land-based ice—are melting also. As they do, the massive quantities of water released are pouring into the ocean, raising sea levels around the world.

"After at least 2,000 years of little change, sea level rose by roughly 8 inches over the past century," the U.S. Global Change Research Program reports. Over the past 15 years, oceans have risen at double that rate as land-based glaciers have melted.

In addition to storing freshwater, Arctic ice—on land or sea—is a kind of global air conditioner, reflecting the sun's radiation away from the Earth and sending cold air and water to other parts of the world. That helps to modulate air, land, and sea temperatures. It also helps to power important ocean currents like the Gulf Stream, which moderates temperatures in Western Europe with warm waters from southern latitudes.

As Arctic ice disappears, those global benefits go with it. When it melts, moreover, Arctic sea ice is replaced by seawater on the Earth's surface. Rather than acting like a mirror, like ice, to reflect the sun's heat away, water acts more like a sponge, absorbing solar heat and retaining it.

"You've got more and more of the sun's radiation staying in the Earth's system," said Parkinson, "which contributes to future warming."

There are those who shrug and rightly note that Arctic sea ice has grown and shrunk before. Fair enough. Scientists know Arctic ice cover retreated around 1,000 years ago, though there's nothing to suggest it was completely gone. And, while there was a similar warm period 8,000 years ago, when it is possible the region was ice free, to point to hard evidence of the last time the Arctic had no ice, we have to look back 65 million years.

"You have to go back to the age of the dinosaurs," said Walt Meier, a scientist with the National Snow and Ice Data Center, a research and data archival partnership among NASA, the National Oceanic and Atmospheric Administration, the National Science Foundation, and the University of Colorado at Boulder.

What we do know for certain is this:

As a regulator of global temperatures, ocean currents, and the availability of water that controls global sea levels, Arctic ice is essential to the world as we know it, and the fact is it's melting at an alarming rate.

Uncertainty, it turns out, can cut both ways. While our foot's been stuck on the gas pedal, we may have gone farther than we know.

"Things are happening a lot more quickly in the Arctic than we had thought," said Meier. "The ice is in a lot more fragile state than we had thought."

Several thousand miles south of the melting Arctic ice, the effects of climate change reveal their force in fire—wildfires, specifically, which are burning, on average, 8 million acres of American land each year. That's double the ten-year average of just a decade ago.

A single wildfire can cause losses in the hundreds of millions of dollars.

The 2002 Rodeo-Chediski fire near Winslow, Arizona, burned nearly half a million acres across two national parks and the Fort Apache Indian Reservation. Damages from lost homes, timber, local businesses, and other losses topped $260 million, in addition to the $50 million spent fighting the fires, according to a study by the Western Forestry Leadership Coalition, a partnership of 33 state and federal forest service officials from Western states.

Wildfires exact a human toll as well. In the first eight months of 2009, 14 firefighters died fighting wildfires in North Carolina, Missouri, California, and elsewhere, according to the International Association of Wildland Fire, a nonprofit outfit based in Birmingham, Alabama, that represents firefighters.

Throughout the 1990s, wildfires in this country burned, on average, 3.8 million acres a year. In four of the five years that ended in 2008, total acres burned exceeded 8 million. Over the next five years, wildfires will burn between 10 million and 12 million acres of American forests and fields each year.

Those statistics and predictions come from the January 2009 Quadrennial Fire Review (QFR), a joint document produced every four years by the U.S. Forest Service, National Park Service, Fish & Wildlife Service, Bureau of Indian Affairs, Bureau of Land Management, and the National Association of State Foresters.

They should know. And they're clear about the reasons why.

"The effects of climate change will continue to result in greater probability of longer and bigger fire seasons, in more regions of the nation," the QFR reports. "What has already been realized in the past five years—shorter, wetter winters and warmer, drier summers, larger amounts of total fire on the landscape, more large wildfires—will persist and possibly escalate."

Hard hit will be woodlands and fields across the Southwest and Southeast, where long-term drought conditions are expected to persist for at least another 20 years.

"Additionally, large, costly, and damaging wildland fires will be more likely to occur in geographic areas that haven't traditionally experienced such events, such as the Midwestern, Eastern and Southeastern part of the U.S.," the QFR warns.

Not only is climate change expanding the range of wildfires, it is extending the wildfire season each year by increasing the kind of warm, dry weather that creates conditions conducive to burning.

"Studies already indicate that wetter and warmer winters followed by faster snowmelt in the West have expanded the fire season horizon," the report states. "Research has also confirmed that fire seasons are lengthening, indicating that 30 days or more should be added to the start of the traditional fire season and possibly the end."

Climate change is taking its toll on U.S. forests in another way.

Between 1997 and 2001, we lost an average of 2.5 million acres of forests a year due to insects and disease. Between 2002 and 2007, though, the average soared to 8 million acres a year. The increase is important. The more trees die of insect attack, the more acreage of dry, dead trees—the perfect fuel for wild-fires.

A major reason for the massive die-off, the QFR reports, is warming temperatures that extend the range of insects that prey on healthy trees.

President Obama got a first-hand look at just one example in August, when he and his family visited our first, oldest, and most beloved national park, Yellowstone. There, amid the wild domain of bison and elk, he witnessed the modern devastation visited upon an ancient tree, the whitebark pine.

The signature species of the northern Rocky Mountains ridge line, the whitebark pine is a foundational tree critical to the high country habitat of red squirrels, elk, and grizzly bears. A slow-growing species that lives for centuries, the whitebark pine is often the only tree hardy enough to withstand the frigid winters and harsh winds at several thousand feet elevation. These majestic trees help to shelter smaller plants, stabilize vulnerable mountaintop soil, moderate snowmelt run-off, ensure steady stream flow in summer, and produce a super-sized pine nut that's essential food for wildlife, particularly Yellowstone's iconic grizzly bear.

After anchoring the Rocky Mountain high country for thousands of years, the whitebark pine is threatened with extinction by climate change, which has warmed the northern Rockies just enough to allow the native mountain pine beetles to flourish at high elevations where few could thrive before now. As a result, as much as 70 percent of these ancient trees are already dead in parts of Montana, Wyoming, and Idaho.

The mountain pine beetle is just one example of a global trend brought on by climate change.

"Already 20,000 data sets show a wide range of species on the move," the World Bank reports, with overall habitat migrating away from the Equator, north and south, by an average of 4 miles per decade, or upward along mountain slopes, again, on average, about 4 feet every ten years, "as an apparent result of the increase in temperatures."

In 1872, just seven years after the Civil War, our Congress established Yellowstone National Park as the first such delineation of public lands in our country— indeed, the first anywhere in the world. From that inspired decision has grown a network of nearly 400 national parks for Americans to enjoy.

Not even lands set aside for posterity, though, are beyond the reach of climate change threats.

"Human disruption of the climate poses the greatest threat our national parks have ever faced," states a September report my organization, the Natural Resources Defense Council, coproduced with the Rocky Mountain Climate Organization, a research and advocacy group based in Louisville, Colorado.

Among the report's findings:

Pinon pines are dying off in New Mexico's Bandelier National Monument, Utah's Zion National Park, and Nevada's Lake Mead National Recreation Area, as heat and drought has weakened the trees, making them more susceptible to attack by the pinon bark beetle.

Glaciers are in dramatic retreat—losing 20 percent to 50 percent of their mass in some places, and disappearing entirely in others—in North Cascades National Park in Washington State, Rocky Mountain National Park in Colorado, and Glacier National Park in Montana. In Alaska's Glacier Bay National Park, the shifting weight of melting ice has led to earthquake fears. In the twenty-second century, people will wonder how the "glacier parks" got their names.

Winter has been shortened by up to a month in Wyoming's Grand Teton National Park, the Indiana Dunes National Lakeshore in Indiana, and the Acadia National Park in Maine, reducing the season for snowshoes and skis.

Coastline—more than 7,000 miles of it—is vulnerable to rising sea levels at 74 National Park System properties. Already parts of Assateague Island National Seashore, straddling Virginia and

Maryland, have been breached by encroaching seas. Much of North Carolina's Cape Hatteras National Seashore is facing a similar threat, as is Padre Island National Seashore in Texas, site of the world's longest undeveloped barrier island.

Our National Park Service may be the best nature management agency in the world. Its facilities draw 275 million visits a year. None of that, and nothing else, has yet spared our national parks from the ravages of climate change, or the threat of more to come. That work is yet to be done.

As global ice retreats, wildfires blaze unabated, and even the natural splendor we've set aside as sanctuary comes under assault, Americans feel drawn to preserve for future generations what others of vision protected for us.

Eight in ten Americans (83 percent, to be precise) favor stricter laws and regulations than those we already have on the books to protect our environment, a statistic that hasn't changed over the past decade, according to polling by the Pew Research Center for People & the Press. Nearly six in ten (57 percent) support specific legislation pending in Congress to address global climate change, according to a mid-August Washington Post/ABC News poll of 1,001 adults nationwide.

#### CHAPTER TWO

Most of us understand, it turns out, the impacts climate change is having on our daily lives—exacerbating droughts across our croplands, dropping water levels in our lakes, increasing the rate of wild-fires, and threatening our drinking water. There are others, though, beyond our shores, for whom climate change means something more: a grinding, daily trag-edy striking at the heart of survival.

# 3

# CRADLE OF PERIL

I t's called the Cradle of Humanity, the Great Rift Valley of East Africa, where remains of our prehistoric ancestors stretch back 4.4 million years. By the fall of 2009, though, Rift residents were eating cactus to survive, as the worst drought in a decade dried up ancient waterways, killed cattle by the thousands, and scorched crops to a withered ruin.

In Kenya, 3.8 million people—10 percent of the population—struggled to get by on as little as a bowl a day of corn porridge.

"Children are on the brink of death," Catherine Fitzgibbon, deputy director of Save the Children in Kenya said in a September press statement. "The number of malnourished children coming to our feeding centers is going up and up," she said. Weakened by malnutrition, they're vulnerable to death by diarrhea, dysentery or malaria, she said. "We expect it to get worse."

Months of raging wildfires denuded thousands of

acres in Mau, the largest forest in East Africa and the watershed that nourishes a dozen major rivers. Water flows to Lake Victoria, headwaters of the Nile, fell off sharply, as rivers flowing from Mau slowed to a trickle or disappeared in a muddy ooze.

Elephants died by the dozens; their sun-bleached bones a searing testament to final days of hunger and thirst. Cattle and goats dropped by scores of thousands, amid estimates that half the country's livestock would perish. Desperate herders drove cattle deep into the Masai Mara National Reserve in search of water, while rangers fought to protect dwindling supplies to keep wildlife alive.

Close to a million Kenyans, whose families have herded goats and cattle for centuries, quit the pastoral lifestyle, unable to make a go of it. As once-lush grazing lands have gone to barren clay, families by the hundreds have repaired to makeshift squatters camps along the road to Mandera.

"These villages are now hosting the very first 'environmental refugees' in northern Kenya," Mohamed Adow, an aid worker with Christian Aid, wrote in the October 2008 edition of *Forced Migration Review*, a quarterly published by Oxford University's Refugee Studies Center. In Kenya, these villagers are called "pastoralist dropouts," wrote Adow. "The way of life that has supported them for thousands of years is falling prey to the impact of climate change."

The cradle of human origins is telling a story of human demise, a grim preview of the catastrophic toll climate change has already begun to impose on our world.

Nor is the peril limited to Africa.

Closer to home for Americans, the summer of 2009 brought the worst drought in 70 years to Mexico, killing an estimated 50,000 cows and wiping out 17 million acres of cropland. A few months earlier it was China, reeling from its worst drought in 50 years, threatening 25 million acres of croplands directly affecting an estimated 4 million people.

Across much of the developing world, in fact, climate change is grinding away daily at the health, welfare, and basic security of millions of low-income people living on the jagged edge of environmental catastrophe. Thousands are already dying each year because climate change is:

- widening desertification, reducing available pasture and croplands, eliminating critical sources of food;
- expanding conditions favorable to the breeding of certain types of mosquitoes that transmit dis-

eases like malaria, which kills more than 800,000 people each year, and even dengue fever;
- exacerbating droughts, contributing to a desperate scarcity of clean water, the lack of which can lead to diarrhea, which kills 2.2 million people, mostly children, every year; and
- increasing the devastation of storms and floods.

Those are the conclusions of the Geneva-based Global Humanitarian Forum, a nonprofit research and advocacy organization founded in 2007 by former U.N. Secretary-General Kofi Annan to highlight the human toll of global climate change.

"The findings of this report indicate that every year climate change leaves over 300,000 people dead, 325 million people seriously affected, and economic losses of U.S. $125 billion," Annan said last May when the group released "The Anatomy of a Silent Crisis," a 136-page survey of the impact climate change is having on poor people around the world.

"Four billion people are vulnerable, and 500 million people are at extreme risk," said Annan. "In the next twenty years those affected will likely more than double—making it the greatest emerging humanitarian challenge of our time."

It's also a moral challenge.

"This is a sin against humanity," said Representative John Lewis, D-Ga., "to see people suffering, dying, for the lack of clean water, or clean food. . . . I happen to believe that it is something very violent. It is not in keeping with the philosophy of non-violence."

A genuine pillar of moral authority, Lewis was an early proponent of nonviolence in the American civil rights movement. He marched alongside the Rev. Dr. Martin Luther King Jr. and bears the scars from those marches in Selma, Montgomery, and Birmingham. An ordained Baptist minister, Lewis still works for justice—social, economic, environmental.

"More and more religious leaders, more and more people of faith, are saying 'This is God's creation, and we don't have a right to destroy it,'" Lewis said in a recent conversation. "We don't have a right to make it almost impossible for people to breathe clean air, or drink clean water or eat safe food."

The Global Humanitarian Forum's estimates assume that climate change is responsible for just 4 percent of the world's most serious environmental degradation. The estimates also blame climate change for 40 percent of the increase in weather-related disasters, the incidence of which has doubled since 1980. Those assumptions are based, respectively, on global health data provided by the World Health Organization (WHO), an arm of the United Nations, and by analyzing extensive weather-related disaster records from insurance companies, aid agencies, and other sources stretching back a full century.

"Overwhelming evidence shows that human activities are affecting the global climate, with serious implications for public health," states a 2008 WHO report. "Catastrophic weather events, variable climates that affect food and water supplies, new patterns of infectious disease outbreaks, and emerging diseases linked to ecosystem changes are all associated with global warming and pose health risks."

More than half of the world's population now lives within 40 miles of shorelines, the WHO reports, leaving them vulnerable to exposure to rising sea levels and to the storm surges that accompany hurricanes, typhoons, and tropical storms. In Bangladesh alone, where one-fifth of the low-lying country is flooded each year, storms and droughts have killed 9,000 people and caused more than $5 billion in damages in just the past nine years.

Additional warming going forward would impact low-income people more directly than those of greater means.

A further increase of just 1.8 degrees in average global temperatures would reduce the availability of clean water to at least an additional 400 million people, and perhaps as many as 1.7 billion, mostly in Africa and Asia, the World Bank estimates in its World Development Report 2010, released in mid-September.

Coral reefs would die—devastating the populations of fish and threatening the people who depend on them for their livelihood—and perhaps a quarter of the world's plant and animal species would be flirting with extinction.

"Latin America and the Caribbean's most critical ecosystems are under threat," the World Bank warns. "The tropical glaciers of the Andes are expected to disappear . . . resulting in water stress for at least 77 million people as early as 2020," the report states. "The most disastrous impact could be a dramatic die-back of the Amazon rain forest and a conversion of large areas to savannah, with severe consequences for the region's climate—and possibly the world's."

The disproportionate impact of climate change on the world's poorest people raises questions of basic equity.

"It is a grave global justice concern that those who suffer most from climate change have done the least to cause it," Kofi Annan said in May.

The United States and other high-income nations produced, on average, 15 tons of greenhouse gases per person in 2005, according to World Bank calculations. That's more than seven times the per-capita rate in low-income countries. And yet, it is low-income people who bear the sharpest risk and most immediate consequences of global climate change.

"This," said Annan, "is fundamentally unjust."

I know what he's talking about. I've heard directly from colleagues across the developing world, lamenting the precious resources they devote to contending with the affects of our carbon emissions, diverting needed money from housing, health, and schools.

As a young woman, I spent two months traveling through East Africa. I was taken by the forests of Rwanda, where Dian Fossey studied the ways of great gorillas. I marveled at the lush, dense jungles of Eastern Congo, and the teeming herds of Serengeti game. Even then, though, pressures were emerging on the waters and lands. Signs of habitat degradation were clear, and I returned home haunted by troubling questions about the human toll of environmental decay.

How dire must conditions become, I wondered, how far must justice fall short, before a man will pack up his family, taking just what he holds in his hands, and leave behind the only home he's ever known for the uncertainty of distant lands?

Shortly before his death of cancer in August, Refugees International president Ken Bacon and his wife, Darcy, established a new Center for the Study of Climate Displacement. It will track the rise of people fleeing the effects of global warming and give voice to their plight. After a career that had taken him to the heart of the problems plaguing the 42 million people around the world who have been driven from their home by disaster or war, Bacon looked into the future in the last days of his life and saw untold numbers more displaced by global warming.

"Climate change will force millions of people from their homes," Bacon said at the time, "and this will pose enormous challenges to an already stressed humanitarian system."

Global climate change is the single greatest environmental challenge of our time. And yet, it is far more than that. It is a humanitarian challenge. It is an economic challenge. It is a national security challenge. It is the great moral challenge of our time.

# 4

## Greenprint

### A Blueprint for Change

All living forms contain carbon. And, in a very real sense, the Earth is a giant battery storing the energy contained in carbon-laced remains accumulated over the past 300 million years. In that time, prehistoric plants and aquatic life have successively thrived, expired, and been transformed into coal, natural gas, and oil, the carbon-rich fossil fuels that provide 85 percent of the world's energy.

Practically all of the fossil fuels ever used have been burned since the dawn of the Industrial Revolution two centuries ago, a mere blink of the eye in geologic time. In that veritable moment in time, we have asked the Earth's atmosphere to choke down carbon pollutants resulting from the combustion of fuels it took 300 million years to create. No wonder the planet is gagging.

Every gallon of oil, every lump of coal, every whiff

of natural gas we burn releases carbon dioxide. Much of it goes into the atmosphere, where it becomes part of the air that covers our Earth.

Carbon dioxide acts as a blanket, holding in the Earth's heat as part of what's called the "greenhouse" effect. Greenhouse gases like carbon dioxide are helpful, up to a point. Too much, though, heats up the planet. That's what's been happening. Too much carbon dioxide from burning fossil fuels.

There are natural processes—breathing, for instance, or the decay of dead plants—that put carbon dioxide in our air. These natural processes are balanced by others like growing plants and food that remove carbon dioxide. Deforestation influences those levels because trees store large amounts of carbon in their trunks, reducing the amount of carbon dioxide in the air. And there are other gases—methane is one—that contribute to warming as well.

Nothing, though, comes close to creating the carbon dioxide concentrations, and the consequent warming impact, as the burning of fossil fuels. These emissions make up about 80 percent of the carbon dioxide added to the atmosphere each year, according to the World Bank.

Today, there are 387 parts of carbon dioxide for every million parts of air in our atmosphere—scientists call that parts per million.

For most of the past 800,000 years, our atmosphere contained between 170 and 280 parts of carbon dioxide per million parts of air. At today's levels, though, our air has more carbon dioxide in it—between 40 and 130 percent more—than at any other time in nearly a million years.

That's the main reason the average global temperature—57 degrees—has risen about 1.3 degrees over the past century, mostly over the past several decades.

Carbon dioxide concentrations are continuing to rise as we pump more and more carbon emissions into our air. At the current rate of world fossil fuel use, carbon dioxide emissions will rise from 29 billion tons in 2006 to 33.1 billion tons in 2010 and 40.4 billion tons in 2030, according to the U.S. Energy Information Administration, an analysis arm of the U.S. Department of Energy.

At those levels of emission growth, carbon dioxide concentrations in the atmosphere would increase enough to raise global temperatures by nearly half a degree every ten years.

Down that road lies growing peril. Not for everyone all at once, but sooner than most of us realize, and in ways that affect us all. In ways that mean hardship and even death, to an increasing number of people around the world.

There is an alternative. We can begin to take our

future seriously enough to reduce the impact of climate change.

To do that, we need to do three things:

- Reduce our global warming pollution.
- Promote alternatives to fossil fuels.
- Help our country make a smooth transition to the clean energy future we need.

By achieving these specific goals, we can help to generate American jobs, reduce our reliance on foreign oil, and create a healthier planet for ourselves and our children.

Congress is debating plans for doing just that—on an urgent basis.

As a recent report by the staff of the House Select Committee on Energy Independence and Global Warming puts it:

> Global climate change presents one of the gravest threats, not only to our planet's health, but also to the United States' economy, national security and public health. Scientists warn that we may be approaching a tipping point, after which it will become increasingly difficult, or perhaps impossible, to halt global warming and its catastrophic effects.

With that in mind, the U.S. House of Representatives passed clean energy legislation in June. Now a similar bill is before the Senate. It's a comprehensive blueprint for a clean energy future, a greenprint, you might say, for change.

This kind of clean energy legislation starts by setting limits, or caps, on carbon emissions.

In 2005, U.S. power plants, factories, buildings, and vehicles emitted about 6 billion tons of carbon dioxide into the Earth's atmosphere, about one-fifth of the world total. Using 2005 as a base, we would cut those emissions over time. Some legislation has proposed cuts of 17 percent, or 20 percent, for instance, by 2020.

The idea is to set a national carbon dioxide emissions limit each year. The cuts would be phased in over time. Companies such as electricity producers, petroleum refiners, manufacturers, and others, would be required to obtain government permits for each ton of carbon dioxide they send out of their smokestacks, a bit like buying a ticket to pollute.

Polluters would buy these permits from the government. Some polluters would be given the allowances for free at first. That way they could use the money saved for public purposes, such as returning funds to affected consumers, investing in energy efficiency or clean energy, and competing in energy-intensive industries against countries with a carbon dioxide cap.

Over time, these free allowances would be phased

out and all polluters would need to buy permits. Doing so would put a price on pollution, giving everyone an incentive to reduce emissions and put clean enterprises on an equal footing with dirty ones—ones that now get to pollute for free. The government would use the money from the sale of these permits—which some have equated to renting the public's "ownership" of the atmosphere—for other public purposes such as clean energy or tax relief.

In addition, to create a fluid and transparent market for these allowances, companies could buy and sell these permits from each other as well as buying them from the government. This "trading" could encourage all companies to reduce pollution even faster. The idea behind this so-called "cap-and-trade" market is to limit total emissions, put a price on pollution, charge the polluters only for what they actually emit, and reward innovators who cut their carbon emissions by allowing them to profit from the reductions they make.

Once again, some of the money raised from the sale of permits would be used to help pay for energy efficiency measures in workplaces and homes, or to promote the development of electric cars and renewable energy sources that make use of wind, solar power, or biomass. Some would go to encourage the use of emerging technologies that reduce carbon dioxide emissions from fossil fuels or capture such pollutants before they leave the smokestack.

Electricity and natural gas companies would get permits to help cushion consumers against rate hikes. And adjustments would be provided to help give certain industries, like steel and cement, time to modernize their operations.

I've heard groups that want to kill clean energy legislation try to argue that it doesn't comport with the economic facts of life; that putting a price tag on pollution might somehow trifle with the workings of our free-market system.

That is complete nonsense. You deserve to know why.

Free markets have one purpose: to communicate the value consumers place on goods and services. That's it. Markets exist so that producers can find out what consumers want and how much they'll pay for it.

To evaluate a product efficiently, consumers need information. We need to know, for starters, how much something will cost. Quality also matters to most of us. Reliability matters too. We might also want to know where something was made. We might want to know how it was made.

When we pick up an item at the grocery store, though, or click to order from an online catalog, we have no idea how much climate change pollution our potential purchase contains. Nobody tells us how much carbon dioxide was released to make that product. There's no label linking the product's carbon footprint to its impact on climate change. And the costs of the pollution, the harm done by that pollution, are not included in the price of the product.

In a market that is truly free, that information would be there. That would allow consumers to consider environmental costs in the purchase choices we make. Putting a price tag on pollution would give us the chance to do just that.

That's how free markets work. And there's nothing more fundamental to the workings of a free-market economy than allowing consumers to say directly to producers, "This is how much your product is worth to me, this is how much I'm willing to pay."

When free markets determine pricing, our economy hums along, creating wealth and opportunity by responding to what we value.

We value fresh air and water. We care about our planet's health. We want to leave it at least as clean as we found it. A free market would let us say so in every purchase that we make. That's exactly what clean energy legislation can help us do by putting a price on the pollutants that cause climate change.

"This approach will ensure emission reduction targets will be met, while simultaneously generating a price signal resulting in market incentives that stimulate investment and innovation in the technologies that will be necessary to achieve our environmental goal," the U.S. Climate Action Partnership (USCAP) wrote in "A Call for Action," its 2007 statement of principles and recommendations.

USCAP is a coalition of some of the country's leading corporations: the Ford Motor Co., Alcoa, Duke Energy, Caterpillar, General Electric, and DuPont. These companies, and two dozen more, have joined with the NRDC and several other environmental groups to push for effective climate change legislation. This group is cultivating common ground around economic and environmental goals.

"The United States can prosper in a greenhouse gas constrained world," USCAP concludes. "In our view, the climate change challenge, like other challenges our country has confronted in the past, will create more economic opportunities than risks for the U.S. economy. Indeed, addressing climate change will require innovation and products that drive increased energy efficiency, creating new markets. This innovation will lead directly to increased energy security and an improved balance of trade."

USCAP reaffirmed its position and called for clean energy action from Congress in a 2009 report, "A Blueprint for Legislative Action."

"Climate change legislation can benefit our economy and energy future," the group stated in the January report. "We need a new vision and policy direction to transition from the technologies and practices we relied upon in the twentieth century to the technologies and practices America will need in the twenty-first century. . . . Urgent action is needed."

Cap and trade is not a new idea. We used a cap-and-trade program to help phase out the use of lead in gasoline a generation ago. And, in 1995, a cap-and-trade system was put into place to reduce the effects of acid rain by cutting emissions of sulfur dioxide from coal-fired generators and plants. Opponents of that plan claimed it would be costly, cumbersome, and ineffective, a lot like what foes are saying today about the pending clean energy legislation.

The skeptics, though, were wrong.

"That program is widely regarded as a success, because it helped reduce emissions from coal-fired utilities at much less cost than had been expected," the bipartisan Congressional Budget Office wrote in a September 16, 2009, report.

In fact, the sulfur reduction program "has become an international model for a cap-and-trade program," concluded an independent 2003 audit by Resources for the Future, a nonpartisan Washington think tank. Acid rain pollution fell 25 percent in the program's first year and has exceeded expectations since then. Cap and trade provided "dramatic cost savings" compared to conventional regulation. And the price of premium, low-sulfur coal quickly fell 9 percent, due to innovation spawned by increased demand. "Compliance has been virtually perfect," concluded Resources for the Future.

Like any worthwhile national objective, reducing carbon emissions calls for national will. Change is never easy and seldom comes without a price. Before we adopt clean energy legislation, we need a clear-headed view of its costs.

Fortunately, we have that. Clean energy legislation has been analyzed separately by three independent agencies: the Congressional Budget Office (CBO), the Environmental Protection Agency (EPA) and the Department of Energy (DOE).

All three came up with slightly different numbers. In each case, however, clean energy legislation would cost the average household about as much as a daily postage stamp.

This is what they found.

Clean energy legislation would cost the average American household $160 a year in 2020, according to the CBO, or right at 44 cents a day. The EPA estimated the average per-household cost at between $80 and $111 per year—or 30 cents, on the high side, per day. And the DOE has set the cost of this kind of legislation at $83 a year by 2030, or 23 cents a day.

Those numbers, of course, only tally the costs of the program. They do not reflect the benefits—better health, improved national security, good jobs that can't be outsourced—that clean energy legislation can bring to all Americans.

Ever since those numbers came out, the opponents of clean energy legislation have come out in force, spending untallied millions on newspaper and television ads doing their best to distort the facts, confuse the issue, and generally mislead the country.

You may have heard, for instance, that this kind of legislation could raise energy prices, especially for gasoline. The American Petroleum Institute, the trade association for the Exxon-Mobil Corp. and other oil and refinery companies, even hired event planners to organize staged demonstrations over the summer to protest clean energy legislation.

These disruptions got a lot of attention, but they won't turn back congressional momentum for clean energy legislation. Our future is too important for that.

Americans haven't forgotten that, as our workers staggered last year into the worst recession since World War II, Exxon-Mobil racked up record profits of $45 billion, while families paid $4 a gallon at the pump for summer vacation gas. That's $45 billion in *profits*, on $460 billion in sales. That's more than the gross domestic product of Argentina. If Exxon-Mobil were a country, it would have the twenty-fourth largest economy in the world, just after, as it happens, Saudi Arabia.

With that kind of financial throw weight, you wouldn't expect Exxon-Mobil to sit out the climate change debate. It hasn't. In just the first six months of 2009, Exxon-Mobil spent $14.4 million lobbying Washington officials on climate change and other matters, according to company filings tracked by the Center for Public Integrity, a Washington nonprofit civic interest group.

The Peabody Energy Corp., and its subsidiaries, reported lobbying expenses of $2.7 million during the same period. Peabody, based in St. Louis, Missouri, is the world's largest privately held coal company. It reported earnings of $960 million in 2008 on sales of $6.6 billion.

In addition, the American Petroleum Institute (API) spent $4.1 million on its own lobbying activities during the first half of the year. That doesn't count what API and its members spent over the summer, urging oil company employees to protest climate change legislation at political town hall meetings nationwide.

During these staged rallies, and in full-page ads in major newspapers, opponents of clean energy legislation have claimed it would raise gasoline prices to $4 a gallon. What they don't want you to know is the source of that number: a study by the Heritage Foundation, a Washington think tank that received $50,000 in Exxon-Mobil funding in 2008. Heritage then produced a study concluding that, yes indeed, gasoline prices would rise by $1.38 per gallon—by the year 2035. Twenty-five years from now! That wasn't on the placards, but it is in the report, if you read down far enough.

It's disappointing, especially when so many of our families are facing tough times, to see that kind of money being spent trying to mislead and confuse the American people on something so vital to our future. The oil and coal companies, though, have a right to be heard. You can bet they have been. They've spent millions of dollars, after all, to make sure of it.

Now it's time for the rest of us to weigh in. Not with millions of dollars, but with something worth even more. Millions of voices. Millions of votes. One voice, one vote, at a time.

Some groups, when all is said and done, want to frighten us. Some just don't want change. And some, sad to say, don't seem to be willing to do their part to protect our planet or to pay their fair share for the damage they do.

We've all got a right to our opinions. We don't get to make up the facts.

So when you hear someone trying to scare you, or trick you with fuzzy math, ask where they're getting their figures. Ask who paid for the study. Ask who you can trust.

And remember this: some of the finest energy economists in the world—at the CBO, the EPA and the DOE—have carefully analyzed clean energy legislation and developed their best good-faith estimates of its costs. Nobody paid them to cook up the numbers. The public interest is all they serve. They reached independent conclusions. And they stand by what they say: clean energy legislation will cost the average American household somewhere between 23 and 44 cents a day.

We can help put our people back to work, reduce our reliance on foreign oil, and create a safer planet for ourselves and our children, for less than the price of a postage stamp.

# 5

# PUTTING AMERICANS
# BACK TO WORK

As recession closed in on the down-at-the-heels town of Vandergrift, Pennsylvania, hard times hit Robin Scott where it hurt. A family man struggling to make ends meet in the heart of the state's troubled steel country, Scott was one of 150 workers to be laid off in the fall of 2008, after a local window factory closed.

"When they shut this plant down," he said, "it was the worst feeling I ever had."

Six months later, in the kind of rust-belt Cinderella story no screenwriter could script, Scott and his co-workers were back on the job. The once-shuttered plant had been bought, refurbished, and inventively reborn to manufacture energy-efficient windows for offices and homes. Half the town turned out to help reopen the doors.

"This is more than just a job to me," Scott said at

a ceremony to herald the factory's rebirth. "It's also doing our small part, in this tiny corner of America, to move this country forward."

A green-collar revolution is remaking the American economy—one job, one paycheck, one hopeful family at a time.

Clean energy legislation can help. It can provide incentives that promote energy-efficiency investments— like energy-saving windows—and help us develop fossil fuel alternatives like wind turbines, renewable power sources like solar panels, and the next generation of energy-efficient cars, homes, and workplaces.

Pointing the country toward a clean energy future is going to be a big job. It will require the skills of all of us—carpenters, metal fabricators, tool and die makers, scientists, truck drivers, software designers, computer engineers, and an array of others—people who work with their hands and their minds.

Combined with the economic stimulus package, the clean energy legislation now before the Senate would create 1.7 million such jobs, according to a 2009 study by the Political Economy Research Institute at the University of Massachusetts.

That estimate is based on $150 billion of investments in clean and sustainable energy use. The actual investment could be much larger. A July report by McKinsey & Co., an international economic consult-

ing firm, showed that the country could profitably in-
vest more than $500 billion in energy efficiency up-
grades over the coming decade and get more than $2
in energy savings for every dollar spent.

We're talking about many hundreds of thousands of
jobs, spread across all 50 states.

Ohio alone could produce almost 70,000 jobs—
opportunities for steelworkers, coil winders, and bear-
ing makers who produce components to build wind
turbines; electricians who install solar panels; and
construction workers who retrofit buildings to cut energy use.

Robin Scott's story makes the point.

Seven years ago, the company that brought Scott's
plant back to life didn't even exist. Now that company,
Serious Materials of Sunnyvale, California, employs
250 people in three locations. Apart from making energy-efficient windows, the company also uses recycled
materials to produce Ecorock—a new generation of
sheetrock that requires 80 percent less energy to make
than traditional gypsum drywall.

Other examples abound:

- In Caledonia, Wisconsin, Calstar Products refit-
  ted an old radiator factory that uses waste ash
  from coal plants to make "green" bricks that re-

quire 85 percent less energy to produce than clay bricks.

- In Menlo, Iowa, SynGest, Inc., is investing $100 million in a plant that will employ 200 people and buy 150,000 tons of local corn cobs each year and turn them into fertilizer that would otherwise be made from natural gas.

- And outside Houston, Texas, Great Point Energy is building a plant to turn coal into cleaner-burning methane gas. Crucially, the plant will capture carbon dioxide produced as a by-product, then sell it for enhanced oil recovery, a way to coax oil out of aging wells that have lost the natural pressure that pushes crude outward. That means more domestic production without drilling new wells in sensitive areas. And at the end of the process the carbon dioxide will be safely trapped underground.

Each of these companies is part of a group called Environmental Entrepreneurs, or E2, which partners with my organization, the NRDC. Together, E2 members represent some $20 billion in investments that have created more than 1,200 companies that already employ more than 400,000 people nationwide.

These are jobs of the future, they're here today, and not a minute too soon. As of October 2, 2009, there were 15.1 million Americans looking for work. That's the highest number of unemployed Americans since at least the Great Depression, according to the Bureau of Labor Statistics. Our country needs the economic shot in the arm that the shift to a clean energy future can bring.

Wedding energy efficiency to economic growth is not some short-term fix. It's part of a larger, more permanent shift, and the long-term gains would be stunning, as McKinsey showed in its July report entitled "Unlocking Energy Efficiency in the U.S. Economy."

McKinsey's year-long study showed that the United States could save a staggering $1.2 trillion in energy costs, reduce energy consumption by 23 percent, and eliminate 1.1 billion tons of carbon dioxide emissions annually by investing $520 billion in energy efficiency improvements over the next decade.

The benefits depend on the country having a plan for an effective transition to a clean energy future.

"Energy efficiency offers a vast, low-cost energy resource for the U.S. economy," the report states, "but only if the nation can craft a comprehensive and innovative approach to help unlock it."

Not only would that help put Americans back to work, it would help lay the foundations for a generation of global competitiveness in the burgeoning world market for energy efficiency and clean fuels.

"The great race of the twenty-first century will be to provide affordable clean energy to the world," said House Select Energy and Global Warming Committee Chairman Edward Markey, D-Ma.

"The Europeans, Japanese and, increasingly, Chinese are using their domestic policies to drive the development of clean energy industries and stake their claims to the burgeoning global clean energy economy," Markey said at a September committee hearing. "If we want to be globally competitive, we must do the same."

I've seen firsthand what's afoot in China. After three decades of leading the world in economic growth rates, the Chinese are out to become the top producer of green products for a waiting world.

Suntech Power Holdings, based in Wuxi, is China's largest manufacturer of panels that turn sunlight into electricity. Cushioned by handsome government subsidies, Suntech is selling its solar panels in the United States for less than it costs to make and ship them, the company's founder told *The New York Times* in August. The next step, Suntech founder Shi Zhengrong explained, is for the company to build manufacturing plants in the United States, much as Japanese automakers have done.

South Korea aims to become the lead supplier of

batteries for plug-in hybrid cars. European Union governments are subsidizing the development of strategic industries—including high-performance electricity grids and high-tech building insulation—in the hope of establishing an early market beachhead in the contest for clean tech supremacy.

"Other nations realize a critical truth," President Obama told reporters at the White House last June. "The nation that leads in the creation of a clean energy economy will be the nation that leads the twenty-first century economy. Now's the time for the United States of America to realize this as well."

Last summer I spoke with laid-off workers in Gary, Indiana, a once-thriving steel town staggering through tough times. Standing in a packed and steamy union hall, though, I heard the voice of the American future.

I'd never stopped to think about it, but it takes 250 tons of steel to make single wind turbine. That means something in the American heartland.

"This is about jobs, jobs, jobs," said Tim Conway, international vice president of the United Steelworkers, representing 1.2 million Americans nationwide. "And this is about leaving a clean environment for our kids."

American workers don't want a handout. All they want is the chance to compete. Clean energy legislation would help to give them that chance. It's a vote of confidence in the American worker, a vote of confidence our workers deserve.

Far from that Indiana union hall, in a San Diego suburb, the clean energy future we need to create is coming off the assembly line, in the form of a hybrid bus that looks like a bullet train.

In early 2009, I got a front-row seat on one of these buses to our clean energy future, in a visit to the company that makes them, the ISE Corp. Its vehicles, company engineers explained, are powered by electric motors. Auxiliary combustion engines—burning diesel fuel or gasoline—run only when needed to generate electricity to bolster the high-performance batteries.

Low emissions and fuel efficiency are the hallmarks of these vehicles. They are not experimental; there are 230 already in use, with more than 10 million operating miles accumulated to date, in cities like New York and Las Vegas. More will be showcased at the 2010 Olympic Winter Games in Vancouver and the London Summer Games two years later.

The American entrepreneur has always been able to look over the horizon of uncertainty and challenge to discover the opportunity and promise beyond. Harnessing that vision to the skills and the drive of our workers has been the key to centuries of American prosperity from one generation to the next.

Now those forces are joining hands once again, setting our country on the path to an energy future that's clean and secure. The comprehensive energy plan now before the Senate can help to get us there.

You may have heard some of the people arguing against clean energy legislation claim that it isn't what our economy needs. One group making this unfortunate claim is the National Association of Manufacturers, the Washington lobbying organization that represents manufacturing companies.

I'm proud of what American factories produce. I know they can compete. Over just the past decade, though, we've lost 5.5 million manufacturing jobs in this country. Good jobs, for good people. Jobs sent overseas, outsourced, downsized, eliminated, as factories were padlocked coast to coast.

Now the same folks who closed those factories want to stand in the way of good legislation that can help put Americans back to work. We're not going to let them get away with it. We believe too much in our workers for that. We believe too much in our future.

"We cannot be afraid of the future," Obama said in the Rose Garden last June. "And we can't be prisoners of the past."

"There's no longer a question about whether the jobs and industries of the twenty-first century will be centered around clean, renewable energy," he said. "The only question is, which country will create these jobs and these industries? And I want that answer to be the United States of America."

Just months after taking office, another American president issued a similar challenge in uncertain times:

> I believe that this nation should commit itself to achieving the goal, before this decade is out, of putting a man on the Moon and returning him safely to Earth.

The president was John F. Kennedy. It was 1961. And while his assassination two years later denied him the chance to see it, the rest of the world watched in wondrous awe that summer of 1969 when Neil Armstrong took the historic first steps on lunar soil that marked the fulfillment of Kennedy's charge.

That stunning achievement was the result of a concerted national effort, made against the backdrop of civil rights struggles at home and Cold War conflicts abroad. The Apollo mission established this country as the world leader in technological innovation, with advances that continue to benefit American industry and families every single day.

Now we are positioned to lead again. We can develop the green energy technology of tomorrow, giving a new generation of American workers a leg up in the global marketplace and creating a brighter future for our country.

We can raise our collective ambitions. We can aim high once again. We can generate jobs, reduce our reliance on foreign oil, and create a healthier planet for ourselves and our children. We must rally around those shared goals.

# 6

# A Nation Confident Strong, and Secure

On the last day of January 2006, Coretta Scott King died. American troops were fighting grinding wars in Afghanistan and Iraq. New Orleans was still a sodden ruin in Katrina's wrathful wake.

It was "a time of testing," President George W. Bush told the nation that night in his State of the Union address.

Through it all, he assured the country, we remained confident, strong, and secure, words he used a total of 16 times in his 51-minute speech.

There was, though, something troubling Bush, something that put the country at risk.

"Keeping America competitive requires affordable energy," the commander in chief declared. "And here, we have a serious problem. America is addicted to oil, which is often imported from unstable parts of the world."

Then, as now, every time an American anywhere puts ten gallons of gasoline in their car, about six come from abroad. The son of a Texas oilman, and a man who once started an oil company himself, Bush understood as well as anyone alive what that kind of overdependence meant for our diplomacy, our economy, and our national security.

With some 140,000 U.S. troops in Iraq, where more than 2,200 had already been killed and twice that many would eventually die, Bush set a national goal: "to replace more than 75 percent of our oil imports from the Middle East by 2025."

It won't be easy, he told us. It won't come without a price.

But we have it within our means, he said, and we owe it to ourselves to harness American inventiveness to meet American needs and develop the green energy technologies of tomorrow.

Bush laid out what he called his Advanced Energy Initiative, a 22 percent boost in federal funding to help develop high-performance batteries for hybrid cars, ethanol fuels from domestic wood chips and corn, and new ways to harness the power of the wind and the sun.

"By applying the talent and technology of America," he said, "this country can dramatically improve our environment, move beyond a petroleum-based economy and make our dependence on Middle Eastern oil a thing of the past."

Today, more than ever, that vision endures, a national promise worth keeping.

Americans spent a record $450 billion on imported oil in 2008. American dollars—$1,400 for every man, woman, and child in this country—sent overseas, to places like Saudi Arabia, Venezuela, and Russia. Half a trillion dollars, in a single year, earned in this country and then sent abroad, beyond the reach of our people. Half a trillion dollars that will never be used to improve our health care, our factories, our schools.

Is that the best thing we can do with half a trillion dollars a year? Does that make us strong and secure?

"Our growing reliance on fossil fuels jeopardizes our military and affects a huge price tag in dollars and potentially lives," retired Navy Vice Admiral Dennis McGinn told the Senate Foreign Relations Committee in July. The U.S. dependence on foreign fuel, he said, "undermines our moral authority in diplomacy and weakens U.S. international leverage, entangles the United States with hostile regimes and undermines our economic stability. In our judgment, a business-as-usual approach constitutes a threat to our national security."

The clean energy legislation that's before the Senate would go a long way to change that. It could cut our oil imports in half.

The bill would promote fuel efficiency and renewable fuels, helping to reduce U.S. oil consumption, the Department of Energy estimates, by more than 1 million barrels a day. The bill also supports a practice called enhanced oil recovery that lets us use carbon dioxide captured from industrial operations to increase our output of domestic oil.

An oil well is a bit like an aerosol can; it can run out of pressure and still have liquid left inside. There are up to 60 billion barrels of oil stranded in aging U.S. wells that have lost their natural pressure, the Department of Energy estimates. We can capture carbon dioxide before it leaves our smokestacks, then inject it as a pressurized gas into stagnant wells to force out remaining oil.

This is not pie in the sky. The technique has been used successfully for decades in states such as Texas, Wyoming, and Mississippi. Clean energy legislation would help expand this promising practice and increase the availability of carbon dioxide gas for this use. We can bury carbon pollutants deep underground and pump out 5 million barrels of domestic oil every day for 30 years.

Combined with efficiency improvements and renewable fuels, that's 6 million barrels a day of imported oil we could replace or do without under the provisions of the clean energy bill. That could cut U.S. oil imports in half. And that would make our country more secure.

Reducing our oil imports, in fact, has been a U.S. strategic goal for more than 30 years. It hasn't worked out very well.

After an Arab oil embargo and OPEC pricing spikes sent our economy reeling and forced fuel rationing, President Jimmy Carter called in 1977 for a national conservation and fuels replacement effort he said would be "the moral equivalent of war."

Two years later he drew a line in the sand.

"Beginning this moment, this nation will never use more foreign oil than we did in 1977—never," Carter said in a televised speech. "This intolerable dependence on foreign oil threatens our economic independence and the very security of our nation."

The year he spoke those words, we imported 8.5 million barrels of oil a day—roughly half of our oil supply. We now import 11.6 million barrels a day, about 60 percent of what we consume. We blew across Carter's line in the sand—then just kept right on going.

"For more than three decades we've talked about our dependence on foreign oil, and for more than three decades we've seen that dependence grow," President Barack Obama said last June. "There's no disagreement over whether our dependence on foreign oil is endangering our security. We know it is."

When presidents as different as Jimmy Carter, George Bush, and Barack Obama agree on the need for strategic change, shouldn't we pay attention to the warnings they sound? After 30 years of talking about reducing our dependence on foreign oil, isn't it finally time to act?

∽

Diary Entry: Sometime Before 2025
Office of the President of the United States

*Many countries have been preoccupied with achieving economic growth at the expense of safeguarding the environment. The scientific community has not been able to issue specific warnings, but worries increase that a tipping point has been reached in which climate change has accelerated and possible impacts will be very destructive. New York City is hit by a major hurricane linked to global climate change. The New York Stock Exchange is severely damaged and, in the face of such destruction, world leaders must begin to think about taking drastic measures, such as relocating parts of coastal cities.*

The scenario above is a work of fiction. It is not, though, some Hollywood pitch. It is taken, verbatim, from a document produced by the National Intelligence Council, the long-range strategic planning arm of the U.S. intelligence community. It is the council's job to look into the future and envision global changes that could threaten the national security of the United States.

In November 2008, the council published a survey entitled "Global Trends 2025: A Transformed World." Climate change, and the adverse affects it is already having on parts of the world, played a central role in the council's vision of the future threat to the landscape.

"We judge global climate change will have wide-ranging implications for U.S. national security interests over the next 20 years," Thomas Fingar, who was then the deputy director of national intelligence—and the chairman of the National Intelligence Council—testified before the House Select Committee on Intelligence. "We judge the most significant impact for the United States will be indirect and result from climate-driven effects on many other countries and their potential to seriously affect U.S. national security interests."

As U.S. dependence on foreign oil has put our national security at risk, intelligence analysts and Pentagon planners have begun to look at climate change in a similar light.

They see a shifting strategic landscape. They see fragile governments struggling to secure clean water and food across a band of volatility stretching from Africa through the Middle East and into Central Asia. They see support systems overwhelmed worldwide by the growing virulence of violent storms. They see people by the scores of millions living on the edge of environmental collapse. They see trend lines showing every one of these risks tracking sharply up. And they know exactly who's going to get the 911 call when it all starts falling apart.

"We will pay for this one way or another," retired Marine Corps four-star General Anthony Zinni wrote in a 2007 report by the Center for Naval Analysis, an independent research group. "We will pay to reduce greenhouse gas emissions today, and we'll have to take an economic hit of some kind. Or we will pay the price later in military terms. And that will involve human lives. There will be a human toll."

Zinni should know. As commander in chief of U.S. Central Command, he oversaw U.S. military operations across the 4.6 million-square-mile region linking the Middle East to Central Asia. U.S. forces have been fighting for eight years—first in Afghanistan and then in Iraq—in the oil-rich region, which is central to the global antiterror campaign.

Zinni, moreover, is far from alone in seeing climate

change as a national security concern. *The New York Times* reported in August 2009 that military planners have been war-gaming scenarios in which desertification, drought, famine, water scarcity, hurricanes, climate refugee flows, and other consequences of climate change impose new demands on U.S. forces and could, in some cases, provide the flash point for regional conflict. The issue is being included in the Pentagon's next major threat assessment, the Quadrennial Defense Review, due out in early 2010.

"The United States is confronted with a cluster of national security threats that arise from our economic and cultural reliance on fossil fuels," Senator Richard Lugar, a Republican from Indiana, said at a July hearing of the Senate Foreign Relations Committee. "We face international crises arising out of drought, food shortages, rising seas and other manifestations of climate change," said Lugar, the committee's ranking Republican, "all of which could lead to conflict."

A witness at the hearing was even more blunt.

"Climate change is a threat to our national security," said retired U.S. Navy Admiral Lee Gunn. "Taking it head on is about preserving our way of life."

After a 35-year career stretching from the Vietnam War to the collapse of the Berlin Wall, Gunn retired from the Navy to become president of the American Security Project, a nonprofit public policy group with a focus on emerging strategic challenges. The organization's advisory board includes Kenneth Duberstein, who served as chief of staff to former President Ronald Reagan; Senator Chuck Hagel, a Republican from Nebraska; and former Pentagon official Richard Armitage, who was deputy to Secretary of State Colin Powell during the Bush administration.

"Changes in the Earth's climate are more evident every day, but the United States has failed to act, alone or with allies, to avoid disaster," the group states in "Climate Security Index," a recent survey of mounting risks.

"Climate change threatens unrest and extremism as competition for dwindling resources, especially water, spreads," the index reports. "State collapse, massive refugee flows, and increased conflict—both between countries and within them—will be more common."

Similar themes are echoed in the National Intelligence Council overview.

"Experts currently consider 21 countries, with a combined population of about 600 million, to be either cropland or freshwater scarce," the council states. "Climate change is expected to exacerbate resource scarcities."

Already such scarcity is playing a key role in unfolding conflicts across much of Africa, including the arid Sudanese province of Darfur, where an estimated quarter of a million people have been killed in what Bush long ago labeled "genocide." Muslim and Christian families are feuding over irrigation and water diversion in Lebanon's Bekaa Valley. Livestock herders and farmers have skirmished over scarce water supplies in parts of Ethiopia.

"The impact of climate change on international security is not a problem of the future, but already of today, and one which will stay with us," the 27-nation European Union executive branch stated in a 2008 policy paper. "Water shortage, in particular, has the potential to cause civil unrest."

More than 200 international water treaties have been negotiated over the past 50 years, a testament to the growing concern nations have over their ability to ensure access to this essential resource. Some of the greatest pressures are on the already volatile Middle East.

"Water systems in the Middle East are already under intense stress," the European Union reports. "Roughly two-thirds of the Arab world depends on sources outside their borders for water. The Jordan and Yarmuk rivers are expected to see considerable reduction in their flows affecting Israel, the Palestinian territories, and Jordan. Existing tensions over access to water are almost certain to intensify in this region, leading to further political instability."

Climate change doesn't have to actually trigger war in order to destabilize provinces, countries, and entire regions.

"Climate change has also been linked to terrorism," warns the Global Humanitarian Forum, "because it can serve as a threat multiplier for instability in the most volatile regions of the world, which are vulnerable to civil unrest and the growth of extremist ideology."

There's a long and tragic history of government collapse, civil unrest, and ultimately bloodshed when large groups of people find themselves unable to secure fresh food, clean water, and fertile farmland to provide for their families. To imagine the strategic consequences of those looming threats is anything but far-fetched. Indeed, it's only prudent.

Preparing for those threats, the American Security Project makes clear, requires that the country take action now.

"The United States would be more secure," the group states, "if we reduced our carbon emissions and persuaded others to do the same."

Jimmy Carter was right three decades ago to call for a national effort to reduce this country's reliance on imported oil. George Bush was right to echo that call three years ago, after things had gotten much worse. And Barack Obama is right to stand up today and insist that we follow through on what his predecessors wisely saw long ago as a gaping U.S. vulnerability.

Now we face a related concern, one our military planners and intelligence analysts have prudently taken to heart. Global climate change puts our security at risk. It threatens our way of life. We will pay for it now, or we will pay for it later, possibly with American lives.

We have what it takes to reduce those risks before we put lives in harm's way. We have an obligation to do so. And we have the means to act.

Clean energy legislation can help us cut our oil imports in half. It can help us reduce the carbon pollutants that are causing our planet to warm. It can stand as testament to the rest of the world of this country's commitment to change. And it can set a clear example for our friends and allies to follow.

That's what it means to address threats head on. That's what it means to lead. And that's how we'll keep our nation confident, strong, and secure.

# OUR WORLD,
# OUR TIME

Imagine a world where deserts retreat and ice is stable and cold. Imagine healthy forests and fields, where wildfires once raged untamed. Imagine disease and disaster diminished, green pastures beneath blue skies. Imagine a world where no child anywhere must walk 20 miles for a drink of water.

This miraculous world is within our reach. We will grasp it together, or not at all. No government, no law, no corporate board can bring it to us. But we'll need all that, and more.

We each have a role in creating this world, through the choices we make each day. But we depend on government, corporations, and the rule of law to help organize our collective resources around our common goals. And they, in turn, need us.

For if we, as consumers, signal our values through the free market, we speak, as citizens, through our government—one voice, one vote, at a time. The time has come to raise that voice and tell our leaders and lawmakers what we believe and where we stand.

Sometimes the hardest part of change is finding the courage to begin. Climate change has been a century in the making. We won't turn it around in a day, nor can we solve all our problems with the stroke of a pen. Clean energy legislation, though, can begin the long process of change and give us the tools to improve as we move ahead. That's what we need from our Congress. Leading is acting, and the time to act is now.

History presents us with moments infused by great challenge, when the world awaits someone to lead. We have arrived at just such a moment. We need global action on climate change.

As Americans, we have the opportunity to take action that will set us on a clean energy pathway. We have an obligation and ability to lead like no one else in the world.

Under the auspices of the United Nations, countries are trying to agree to a global way forward on climate change. A key meeting is set for December 2009 in Copenhagen, Denmark, where the U.N. goal is to craft a broad accord that will guide each country in setting specific policies. It's an enormous challenge to get 192 nations to agree on a global strategy.

"Copenhagen needs to be the most ambitious international agreement ever negotiated," former U.N. Secretary-General Kofi Annan said in May 2009. "The alternative is mass starvation, mass migration and mass sickness. If political leaders cannot assume responsibility for Copenhagen, they choose instead responsibility for failing humanity."

There are, to be sure, complications.

Developing nations, such as those across much of Africa, Latin America, and Southeast Asia, blame climate change on wealthy countries like the United States and the members of the European Union. Why, these poor countries ask, should they bear the burden for problems they didn't create? Isn't it only fair, they ask, that the countries responsible for climate change chip in to help poorer countries cope with the ills it creates?

Some rapidly emerging economic powerhouses—like China and, to a lesser extent, India, Mexico, South Korea, and Brazil—are taking initial steps while asking for additional time to adjust to the need to limit carbon emissions. After all, they point out, the U.S. and European economies developed without such constraints.

And, finally, advanced countries like the United States and its allies point out that no single country, or group of countries, can solve climate change on their own. What good is it, some wonder, for Americans to cut their carbon emissions, only to see smoke-stack industries across the developing world cancel out those reductions with increased emissions?

These are questions that deserve answers, and there are answers to them all.

Poor countries need help in adapting to the hardships climate change has already imposed. Fast-growing countries need assistance as well, and they can afford to pay for it. They're also the fastest-growing carbon emitters in the world. Carbon dioxide emissions from the United States and other developed countries are on track to increase at 0.3 percent a year between now and 2030. Emissions are set to grow by 2.2 percent a year in developing nations overall, led by China with India a distant second, the DOE's Energy Information Administration (EIA) reports.

There's a bigger picture, though. The global market has created advantages for all of us that didn't exist a few decades ago. The world has gotten smaller, for better or for worse. No country is an island, no economy stands alone. Rich country or poor, cooperation with others, we've learned, is part of the price we all pay for the benefits of global trade. So, too, we all face global challenges, and together we can curb climate change.

"No nation, however large or small, wealthy or poor, can escape the impact of climate change," President Barack Obama told the United Nations in September. "No one nation can meet this challenge alone. . . . Our planet's future depends on a global commitment to permanently reduce greenhouse gas pollution."

It's not easy to make these arguments in a constructive way, or to broker the difficult accommodations that must flow from them. Doing so requires the kind of credibility Americans have painstakingly built over decades spent carefully balancing U.S. interests with those of our global allies, rivals, and friends. That's the kind of leadership we have the opportunity to contribute right now.

"This is precisely the kind of diplomatic impasse that has been resolved, time and again, by concerted U.S. leadership," said career American diplomat Todd Stern, who is Secretary of State Hillary Rodham Clinton's special envoy for climate change. Stern appeared September 10, 2009, before the House Select Energy and Global Warming Committee. In his testimony, Stern made clear that U.S. leadership begins with passing the clean energy legislation pending in Congress.

"What do other countries, both developed and developing, have a right to expect from us? I think, frankly, that we stand and deliver," Stern said. "Nothing that the United States can do is more important for the international negotiating process than enacting robust, comprehensive energy and climate legislation as soon as possible."

We're not doing this for the rest of the world; we're doing it for ourselves. We're doing it because clean energy legislation will help put Americans back to work, reduce our reliance on foreign oil, and create a healthier planet for ourselves and our children. In doing all of that, though, we'll show our global partners what needs to be done and demonstrate the political will to do it and we will be the stronger for having done so.

By acting now, we will preserve and enhance the global competitiveness we need to keep our economy strong. We will strike a blow for the energy independence we need to keep our country secure and we will reaffirm the leadership position the world has come to expect from the United States.

"We have a choice," President Obama said at the White House last March. "We can remain the world's leading importer of foreign oil, or we can become the world's leading exporter of renewable energy. We can allow climate change to wreak unnatural havoc, or we can create jobs preventing its worst effects. We can hand over the jobs of the twenty-first century to our competitors, or we can create those jobs right here in America."

We have the opportunity to lead. It's an opportunity we must embrace.

We also have an obligation to lead.

The United States is the most productive country in the history of the world. In 2008, we kicked out $14.4 trillion worth of goods and services, one-fourth of all the economic activity in the world. In second place came Japan, with $4.9 trillion of gross domestic product, followed by China, at $3.9 trillion, then Germany at $3.7 trillion. When people around the world need help, they tend to reach out to us. More often than not, we respond to the call: ending ethnic killings in the Balkans, dispatching an aid flotilla to Southeast Asia in a tsunami's wake, or sending doctors and dollars to fight AIDS in Africa.

For those reasons, and more, what we do matters. Because we cast such a long shadow across the face of the world, others pay attention to what we do.

We've come a long way, over the decades, in improving our stewardship of the planet. When it comes to carbon emissions, we've got a long way to go. In 2008, EIA figures show that the United States generated 5.9 billion tons of carbon dioxide, roughly 20 percent of the world's total. We're essentially tied with China for the title of top climate change polluter in the world. The world expects better from us than that. I believe we expect more from ourselves.

"A great nation, a powerful nation like America, we have an obligation to get out in front and be a leader in helping others get on the right side of this issue," Rep. John Lewis, D-Ga., said in a recent conversation. "We've been more than lucky, we've been blessed," he said, "and we do have an obligation to do what we can to stop abusing the planet and save it, preserve it for generations yet unborn."

Others contend we have an obligation to lead, if for no other reason than self-interest.

"U.S. leadership alone will not guarantee global cooperation. But if we fail to take action now, we will have little hope of influencing other countries to reduce their own harmful contributions to climate change, or of forging a coordinated international response."

That's the position taken by former Senate Majority Leader Howard Baker, a Republican from Tennessee; former Senate Armed Services Committee Chairman Sam Nunn, a Democrat from Georgia; former CIA Director James Woolsey; former House Foreign Affairs Committee Chairman Lee Hamilton, a Democrat from Indiana; former Defense Secretary William Perry; and two dozen other senior American statesmen in a September statement of policy. It ran as a full-page open letter in *Politico*, a popular Washington newspaper.

"Congress, working closely with the Administration, must develop a clear, comprehensive, realistic and broadly bipartisan plan to address our role in the climate change crisis," the leaders wrote. "We must lead."

As Representative Edward Markey, a Democrat from Massachusetts, put it during a September hearing, "The world is watching the United States."

Someone else is watching, too: a new generation of Americans who, like every generation before them, are asking questions about what we value as a nation and who we are as a people. They are not listening to what we say for the answers, they are watching what we do.

They understand that it is our actions, in the end, that define us and make us who we are. They know our actions will be our legacy, and they will hold us to account. And they're counting on us to make good on the promise of responsible stewardship that every generation, in every time, owes to the next.

"It's like health care and Social Security reform, it's a problem that's going to burden our children," Michael Desch, political science department chairman at the University of Notre Dame, in South Bend, Indiana, said of global climate change. "Is that the legacy we want to leave our children or our grandchildren?"

World leadership is something this country has the ability to provide. It's something we know how to do.

We led, after all, during World War II, sending our sons and daughters to every corner of the globe to turn back a totalitarian movement of horrific proportions and evil designs.

We led during the Cold War in an enterprise of global diplomacy and struggle to defeat a Soviet system that exported repression while imposing misery at home.

And we have led, and we continue to lead, a decades-long democratic, economic, and technological revolution centered around the mightiest political proposition of all time: that free people working in free markets where ideas are free to flourish or to fail is the best way to tap into the boundless potential that resides in the heart and mind of every child ever born anywhere in the world.

This country knows what it means to lead. We must find it within us to lead now.

There are people who care deeply for this country and its future and yet refuse, for whatever reason, to acknowledge the damage before our eyes and the dangers gathering each day.

"Global warming is probably the greatest hoax ever perpetrated on the American people," Senator James

Inhofe, a Republican from Oklahoma, told the Fox news network last June. "The thing is phony."

Inhofe is no political piker. His views can't just be dismissed. He's been in the Senate for 15 years and he's the ranking Republican on the Environment and Public Works Committee and, give him this, when asked how clean energy legislation will fare in the Senate, Inhofe doesn't mince words. "It's dead on arrival," he told Fox. "There's no question about that."

Good Oklahomans sent Inhofe to Washington. When he speaks, he speaks for them.

He doesn't much represent, though, the world as it is, what we see, what we know, what we're told.

We can see that our planet is warming. It's right before our eyes.

The Arctic ice cover is melting. Wildfires burn out of control. Ancient trees are dying. Storms grow more violent. Deserts spread. Clean water gets scarce. Farmland turns to dust. Right before our eyes.

We know it's already harming people around the world—heat wave victims, people who can't feed their families, refugees forced to abandon their homes. We're told 300,000 poor people will die this year, next year, and the year after that due to deprivation and disease exacerbated and even fostered by worsening warming worldwide.

We've been warned of the dangers by people we trust, at the Pentagon, the Commerce Department, the American Academy of Sciences, NASA. The men and women who defend our country, safeguard our prosperity, advance American interests around the world, and expand the domain of human knowledge every single day. The people who put a man on the moon.

Are all these institutions wrong? Are they conspiring to perpetuate some kind of hoax? Is there something these people don't know that our friends at the National Association of Manufacturers, the Exxon-Mobil Corp., and the American Petroleum Institute have somehow managed to figure out?

We all seek American prosperity, because we all care, and care deeply, about American jobs, competitiveness, and the living standards of our families. That's precisely why we need to support clean energy legislation that promotes the kind of research and investment that will help get our workers back on their feet and sow the seeds for future growth in the green technologies that will transform our world.

This is about sound economics. This is about legislation that will create up to 1.9 million jobs, economists at the University of California at Berkeley reported in September 2009. And that's just the beginning of what lies ahead once we begin moving in earnest to a future of clean and sustainable energy in a way that strengthens our economy and our national security.

"Today America is confronting three interrelated crises: an economic crisis, a climate crisis and an energy security crisis," John Doerr, a partner at the venture capital firm Kleiner Perkins Caufield & Byers testified last January before the Senate Committee on Environment and Public Works. "Our best response to all three is a bold, coordinated campaign of investment and incentives to accelerate green innovation. And, in doing so, to ensure America becomes the worldwide winner in the next great global industry: green technologies."

This, as we've seen, is not the stuff of fantasy. Hundreds of thousands of Americans are already on the job, helping to develop and produce the next generation of renewable power sources, alternative fuels, and energy-efficient cars, homes, and workplaces. American entrepreneurs and workers are second to none anywhere in the world. But they must have the tools to compete. We need clean energy legislation to help give them those tools.

The fact is, there are powerful voices astir in the land, just as there always have been, that want things to stay as they are. The status quo is working out pretty well for these folks, and they simply refuse to change.

These groups have their interests to look after, and

they have done that. They have the right to speak out and they have been heard. That's how our democracy works.

Now it's up to the rest of us to do the same. To stand up for our values and give voice to our views. There is more at stake here than self-interest, narrowly defined, for a vocal and powerful few. There's a much larger majority of Americans who understand climate change, see its adverse impact, and want to do something about it. To be heard, we must weigh in. That's how every worthwhile cause in our history has advanced. That, too, is how American democracy works.

"That demand, that moral demand, must be there. And all of the great religions of the world, all of the great faiths, must be part of that effort, to get government, and also the private sector, to move in that direction, in the same way that we did during the civil rights movement," said U.S. Representative John Lewis. "The people have to rise up, along with the community of nations, saying 'This is the direction we are going to move, and we must move. If not, we will not survive.'"

Americans have an innate distrust of received wisdom and the official line. As children of revolution, it's in our blood.

There's a difference, though, between healthy skepticism and an abject refusal to face the truth. We've not come this far as a nation with our collective heads in the sand. We haven't become the people we are by following a misguided but outspoken few who would have us believe a problem as perilous as climate change might simply be wished away. And we cannot allow good governance to be held hostage by those who dismiss danger growing before our eyes as the product of some ill-begotten hoax.

No great country can hazard its future on the views of a radical fringe. That's what this small and shrinking minority has become.

Those who continue, against all evidence, to deny the dangers of climate change are not only at odds with the health of our planet, they are also at war with the essential knowledge of our time.

They have abandoned all pretense of fidelity to fact and retreated beyond the reach of reason to cling stubbornly to yesterday at tomorrow's expense. And they have asked the rest of us to do the same.

They have turned their backs on the greatest scientific minds of our age, the keenest intelligence analysts in our land, and the closest allies we have anywhere in the world. And they have asked the rest of us to do the same.

They have staked their claims on wishful thinking,

contorted logic, and a zealous insistence that the sanctity of profits and ease of old habits can somehow justify environmental degradation and humanitarian suffering on an unconscionable scale. And they have asked the rest of us to do the same.

In the decades, though, since early climate sentries first began to sound the alarm, the gathering weight of evidentiary truth has shifted the center of civic gravity.

We now know, all of us, that our world is warming. We know it is harming us all. We know it is only going to get worse until we stand up and summon the will to stop it. And we know what it will take to do that. We must find the courage to begin.

"The security and stability of each nation, and all peoples, our prosperity, our health and our safety are in jeopardy. And the time we have to reverse this tide is running out," President Obama told the United Nations in September. "And yet, we can reverse it."

Our best history has been written in those times when Americans of all backgrounds, interests, leanings, and means have united around common cause. This is a cause that unites us all.

We all want American prosperity. We want a nation strong and secure. We can see the grim future before us, unless we are moved to act. We see it in our oceans, we see it in our land, we see it in our everyday lives, and we see it, most telling, in the hardscrabble strife of the most vulnerable people on Earth. Those who through no fault of their own find themselves in a life-and-death struggle each day, searching for water, scrounging for food, staving off the ills of disaster and disease for themselves and their families any way they can, and, all too often, coming up short in a fight for their daily survival.

This is the world as we find it. It must not be the world that we leave.

We have what it takes to do better. We have what it takes to lead. A miracle lies within our reach. The moment has come to grasp it.

# EPILOGUE

## One Citizen's Appeal

In American democracy, every voice counts—but only when we speak up.

Let your U.S. senators know why you support clean energy legislation. Respectfully urge them to do the same. They're hearing from the opponents of clean energy legislation. Now they need to hear from the rest of us. That's how our democracy works. It's the only way it ever works well.

Write a letter to the editor of your local newspaper. A lot of people read those letters. Your voice can help them understand this issue.

And talk to family, friends, and coworkers. Don't worry about winning them over. Just get a dialogue going. It's interesting when you stop and think about all that this issue touches and all that we have at stake. You could suggest they take a look at this book. Heck, you could give them your copy—once you're finished of course!

Sometimes the starting point for change is simply asking questions about what we believe or why we do the things we do. Chances are, folks closest to you respect your views and pay attention to what you say. You might be surprised how much influence you have.

Become a citizen advocate for clean energy legislation and you can have a say in what we do as a nation about global climate change. For more information on how to make your voice heard, visit our website at www .nrdc.org. I look forward to hearing from you.

More than two centuries ago, Philadelphia essayist Thomas Paine changed the world by raising his voice and calling on all American Patriots to do the same. Together, we can follow in the footsteps of our nation's founders by heeding their call to raise our voice.

Thanks for reading this little book. I hope you'll share it with a friend. Then help us change the world—again.

# ACKNOWLEDGMENTS

The movement to solve global warming is deep and broad, and I have learned a great deal from those who have added their voices to this fight. Their experience informs these pages, yet I would like to single out a few whose expertise and inspiration made this book possible.

John Adams, the founder of the Natural Resources Defense Council, had an early vision on the critical nature of global warming that propelled NRDC into action early on. Gus Speth, former dean of the Yale School of Forestry and NRDC cofounder and trustee, whose impatience and drive to solve the greatest challenges of the globe have driven us to find solutions and inspired us to work harder. David Hawkins has led NRDC's climate team for many decades with his great intellect and supreme strategic sense. And Dan Lashof and the entire NRDC Climate Center have worked tirelessly to cap carbon and advance clean energy solutions.

Bob Deans helped bring this book to life with his clean prose and thorough research.

I am also grateful for all NRDC staff members and our creative and supportive Board of Trustees, past and present, for committing themselves to our mission to safeguard the earth, and for NRDC's passionate members, without whom our work would not be possible.

Finally, I especially want to thank my husband Paul and three daughters Carrie, Mary, and Lizzie for generously sharing me with the climate fight.

# About the Authors

**Frances Beinecke** is the president of the Natural Resources Defense Council, a national environmental advocacy group guided by its commitment to sound science, the rule of law, and the public interest. In three and a half decades working with the NRDC and representing its 1.3 million members nationwide, Beinecke has established herself as one of our country's most authoritative voices on the need to curb global climate change, a message she has taken from the halls of Congress to the boardrooms of America's leading corporations and to the ministries of the Chinese government.

Beinecke started at the NRDC in 1974, when environmental law and public policy were in their infancy. She began with a focus on preserving the forests of her beloved Adirondack Mountains, later turning to protecting marine life from offshore drilling. She became president of the NRDC in 2006, succeeding its founder, John Adams.

Beinecke earned a bachelor's degree from Yale College and a master's degree from the Yale School of Forestry and Environmental Studies. She has three daughters and lives in New York City with her husband Paul.

**Bob Deans** spent three decades as a journalist before becoming director of federal communications for the NRDC in Washington in 2009. He is a former president of the White House Correspondents' Association and author of the nonfiction book *The River Where America Began: A Journey Along the James*, published in 2007 by Rowman & Littlefield. He lives in Bethesda, Maryland, with his wife, Karen, and their three children.